MAKING THE MOST OF

Light and Mirrors

Linda Thornton and Pat Brunton

Published 2009 by A&C Black Publishers Limited
36 Soho Square, London W1D 3QY
www.acblack.com

ISBN 9781906029753

Cover photos: Jo Wise and Reflections Nursery, Worthing *(top left)*; Shutterstock *(top right)*; Oak Tree Nursery and My Start Children's Centre, Ilfracombe *(bottom)*.

A CIP record for this publication is available from the British Library.

Printed in Great Britain by Latimer Trend & Company Limited

This book is produced using paper that is made from wood grown in managed, sustainable forests. It is natural, renewable and recyclable. The logging and manufacturing processes conform to the environmental regulations of the country of origin.

Contents

Acknowledgements

Our thanks go to the staff, children and parents of:

Reflections Nursery, Worthing, West Sussex

and

Ilfracombe Children's Centre, Devon

Sure Start Chy Carne, Penwith, Cornwall

St Breock Primary School, Wadebridge, Cornwall

Whipton and Beacon Heath Children's Centre, Exeter, Devon.

Photographs taken by alc associates, staff at Reflections Nursery, Jo Wise, Stefano Sturloni and Reflections on Learning.

Introduction

The *Making the Most of* series has been specially devised in order to share good practice, showing what high quality learning and development for young children looks like in real settings. The scenarios described in this book all focus on the use of light and mirrors to demonstrate the exciting opportunities which arise when practitioners observe closely children's self-initiated play and look for the 'extraordinary' in the 'ordinary'.

The role of the practitioner in these child-initiated learning experiences was to provide interesting and unusual resources for the children to explore. The children were then given time to be creative, to try out their ideas, satisfy their curiosity and to become absorbed in the 'serious business of play'. Practitioners paid close attention to what the children were doing and saying and documented their learning in photographs, written observations and transcripts of the conversations. This provided a wealth of information on which to base individual assessments of where to take each child next in order to consolidate and extend their learning and development.

By reading about the scenarios described in this book it is hoped that practitioners will:

- Consider new ways of building on the ideas and interests of the children that they work with
- Extend the range of resources and equipment available to young children
- Explore the use of photographs and transcripts of children's conversations as the basis for planning what opportunities to offer children next

Each of the scenarios has been presented within the context of one of the six areas of learning in the Early Years Foundation Stage (EYFS) and relevant statements taken from the Practitioner Guidance (including the Principles into Practice cards) have been included throughout. However, young children take a holistic approach to learning and it is easy to see that in any one scenario there are connections to many different areas of learning. By the same token, although connections have been made in this book to the EYFS, the statements about children's learning and development are applicable to any curriculum guidance or framework.

How this book is structured

The book is divided into two sections: Section 1 looks at using light and light boxes and Section 2 looks at using mirrors. Each section consists of two different scenarios with accompanying full-colour photographs and descriptive text, for each of the six areas of learning in the EYFS.

- Personal, Social and Emotional Development (PSED)
- Communication, Language and Literacy (CLL)
- Problem Solving, Reasoning and Numeracy (PSRN)
- Knowledge and Understanding of the World (KUW)
- Physical Development (PD)
- Creative Development (CD)

The scenarios illustrate how the principles of the EYFS are put into practice in real life situations. The text has been kept deliberately short and succinct to encourage the reader to focus on the photographs and think about the learning and development that they can see taking place.

Each scenario has the following features:

In the EYFS: Links to specific aspects of the EYFS, including Development Matters

Starting points: A description of the context and the reclaimed resources used

Learning and development: A sequence of photographs and text that demonstrate good practice in the early years linked directly to the statements in the EYFS

Other things to try: Additional ideas on the same theme that you can use to provoke further investigation and development

Creating the right environment

Playing with light and reflection creates magical experiences that can be enjoyed by children and adults alike. Light and darkness hold fascination and intrigue, as well as an element of risk and challenge and of being scared. Experiences involving light and shadow enable children to appreciate the awe and wonder of the world around them and provide an environment that is rich in possibilities for them to develop their natural curiosity. Such opportunities allow adults and children to engage in sustained shared thinking and to develop relationships where joys and fears are open for discussion.

Mirrors can help children to be aware of themselves and to develop a sense of identity. Young children enjoy looking at their reflections and seeing the reflections of others. They are curious about the reflected images they see and want to touch them to figure out whether or not they are real. Children and adults can have interesting and amusing conversations with each other using a mirror, with plenty of opportunities to mimic expressions and movements.

Resources for the exploration of light

Light is best explored in the dark! You can create special places in your setting by adjusting the light levels where light boxes and overhead projectors are used. You could darken small room using blinds or use a naturally dark place such as a cupboard space under the stairs on in a den made especially for the purpose out of blankets and blackout materials.

Light boxes add interest to any setting, creating a place for careful observation or for exploring pattern, shape, form, colour, opacity and colour mixing. The calming influence of a light box invites sensory exploration and provides motivation – engaging attention for sustained periods of time.

By their very nature, the materials that are used to explore light promote aesthetic awareness and an appreciation of beauty. It is equally important that these delicate materials and resources are displayed in such a way that they invite exploration and investigation.

Use divided trays or attractive boxes next to the overhead projector or light box to make the following items accessible:

- translucent materials such as plastic shapes and sheets, buttons, cocktail stirrers, Christmas decorations, small plastic and glass containers and bottles
- opaque materials, including plastic and metal washers, discs, nuts and bolts, lolly sticks, paper clips and coins;
- items with holes in them, such as tea strainers, mesh lids and small strainers, which will give interesting effects
- fabrics, scarves, ribbons and lace, which will have different effects when used with an overhead projector (many of them not what we would expect) providing exciting opportunities for discussion and language development
- natural materials, which can be brought into sharp focus on the light box or overhead projector – use cones, leaves and skeleton leaves, twigs, pods, shells and pebbles

Resources for the exploration of mirrors

Mirrors placed at an angle to each other, such as in a mirror cube, mirror exploratory or a kaleidoscope mirror, produce interesting multiple reflections. Fantasy worlds can be created using mirrors and small world play resources.

Multiple mirrors provide opportunities to develop thinking and problem solving skills, through working out which way to move things to create the desired effect. Moving objects around in front of a mirror helps children to develop their spatial awareness and their early understanding of position.

Resources suitable for use with health and safety approved acrylic mirrors are those without sharp edges that could scratch the surface of the mirrors. Different resources are appropriate for different age groups of children. Resources that can be used safely include:

- paper, card and children's drawings
- fabrics and soft plastics, such as bubble plastic
- cardboard tubing, string and wool
- soft foam building blocks and 3D shapes
- wooden vehicles with rubber tyres
- plastic animals, sea creatures and dinosaurs
- flowers, leaves and skeleton leaves
- cones and seeds without sharp edges

Developing children's creativity and critical thinking

The EYFS *Principles into Practice* states that:

> *'When children have opportunities to play with ideas in different situations and with a variety of resources, they discover connections and come to new and better understandings and ways of doing things. Adult support in this process enhances their ability to think critically and ask questions.'* (EYFS PiP 4:3)

Exploring light and reflection provides the opportunity for you to observe children's investigations, to engage in shared conversations and thinking, and to encourage children to ask questions.

You can support children's enquiry and questioning skills by:

- providing an environment conducive for children to ask questions
- valuing children's answers
- giving children time to think, to formulate questions and to respond to questions that are asked
- modelling a questioning mind yourself

It 's important to remember that many children will not use words to ask questions – some are too young to verbalise their questions and others may lack the communication skills necessary. Children demonstrate their curiosity in many different ways: by gesture, stance and posture; by the length of time they focus on their investigation and by the depth of concentration that they demonstrate.

Supporting active learning

The EYFS *Principles into Practice* states that:

> *'Children learn best through physical and mental challenges. Active learning involves other people, objects, ideas and events that engage and involve children for sustained periods.'* (EYFS PiP 4:2)

Investigating light, shadow and reflection, and using mirrors, are all wonderful ways for children to make connections in their thinking and understanding across all areas of learning. Children can explore these resources as quiet solitary activities or enjoy the excitement of a sharing the experience with others.

As young children explore the equipment and resources, they will develop a wide range of skills that are important for active learning to take place:

- reasoning and thinking skills, including questioning, speculating, inferring, problem solving, recognising similarities and differences, and reflecting
- communication skills of speaking, listening, discussing and recording
- social skills of co-operation, negotiation, following instructions, behaving safely and leadership
- practical skills, including observation, using all the senses, manual dexterity, fine motor control and hand-eye co-ordination

A light box or a mirror exploratory will provide a creative learning environment where children can arrange and sort materials, discover their properties and look carefully at their details. Spending time looking at the details of natural materials, cogs, buttons, acrylic and glass pebbles holds children's attention and creates a sense of wonder about the natural and man-made world around them. By using mirrored equipment, children can develop their mathematical understanding of large numbers, position, repeating patterns, symmetry and counting.

Reclaimed sheets of acetate and clear plastic can be drawn and written on, and illuminated. Coloured translucent plastics can be overlapped and colour mixing experienced. Small world play scenes can be created on light boxes or on mirrored equipment, encouraging children to use open-ended materials in their imaginary play.

Projecting images on to a wall, ceiling or even each other is a joyous opportunity for free creative play as well as providing an ever-changing display. Free exploration and investigation quickly develops into discussion about the properties of different materials (transparency, opacity and so on), colour, shape, form and position. Working with a partner or in a small group also provides the opportunity for co-operation, negotiation and teamwork.

Including a light box in your setting and making good use of an overhead projector will widen the opportunities you can provide for the children to broaden and use their ICT capabilities. Exploring shadows outside at all times of the year will lead to children developing their own theories and ideas about the wonder of light in the world outside.

Health and Safety considerations

Introducing 'new' pieces of equipment into a setting, particularly if these are very different from those traditionally associated with young children, may cause concern for some practitioners. Safety considerations are of course paramount, but there is nothing inherently dangerous about the materials and equipment themselves, provided they are used safely.

Light boxes should be CE marked to comply with IE safety standards. Home made light boxes are not recommended unless they undergo the same rigorous safety checks that are required for electrical goods used in educational establishments. All electrical equipment should undergo an annual portable appliance test (PAT) carried out by a qualified electrician. Liquids should not be used on the surface of light boxes unless they are in tightly sealed containers.

Older style overhead projectors with a bulb in the base unit don't get hot and therefore are safer to use with young children (see 'Other information'). New style overhead projectors are **not recommended** as the bulb is accessible in the top unit and becomes very hot very quickly.

For safety reasons, leads from equipment should be secured to the floor and children should be regularly reminded not to look directly into the light of the overhead projector.

Most Local Authorities do not approve of mirrored equipment that uses glass, for health and safety reasons. In order to stay within health and safety guidelines, it is recommended that acrylic mirrors should be used in Early Years settings. Any wall-mounted mirrors should be installed following the manufacturer's safety instructions.

Section 1

USING LIGHT AND LIGHT BOXES

Playing with light

In the EYFS

The following statements are taken from the Practice Guidance for the EYFS, Personal Social and Emotional Development (Dispositions and Attitudes).

- **Encourage children to explore and talk about what they are learning, valuing their ideas and ways of doing things.**
- **The emotional environment - When children know that their feelings are accepted they learn to express them, confident that adults will help them with how they are feeling.**
- **(Children) display high levels of involvement in activities.**
- **(They) persist for extended periods of time at an activity of their choosing.**

Starting points

Providing a rich and varied environment supports all areas of children's learning and development.

The use of light, both natural and artificial light, can play a major part in establishing a supportive physical environment and in creating the ambience for an appropriate emotional environment.

Including different light sources, such as an overhead projector or a light box, in a setting can suggest a wide range of challenging experiences and activities both to practitioners and to children. Light boxes have a calming influence which invites sensory exploration and provides motivation, sustaining children's attention for long periods of time.

Learning and development

Children react differently to new experiences. As practitioners note how children react to new experiences and activities, they will be able to understand that for some children experiencing light and dark can be both exciting and challenging.

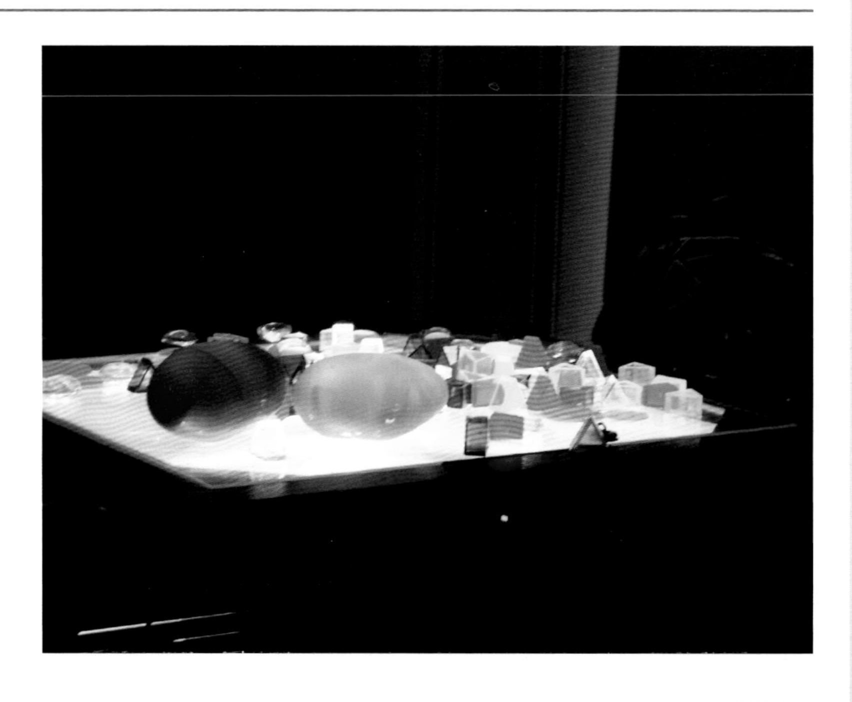

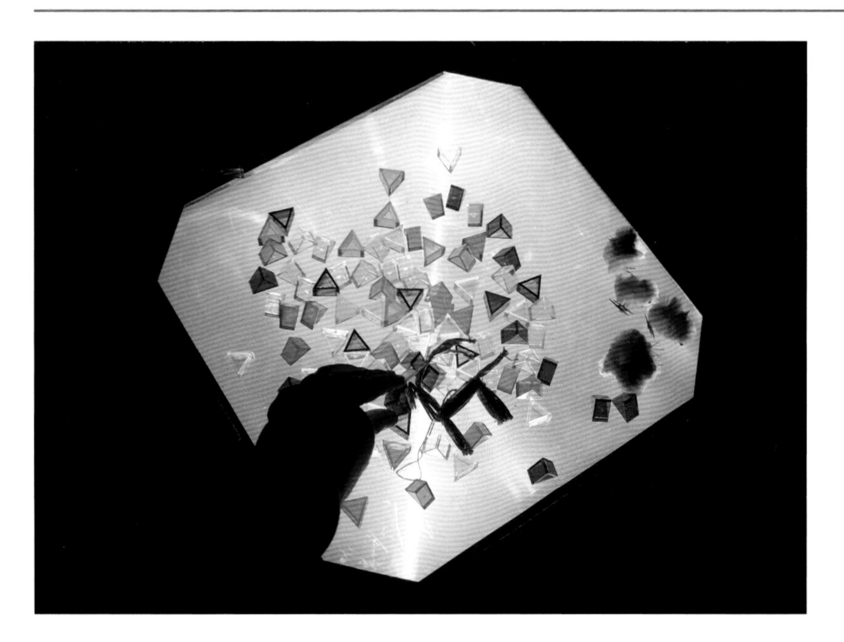

The four year olds in a nursery become absorbed in exploring resources on a lit surface, creating imaginary environments and characters to play out their thoughts and ideas.

They play quietly together, talking about what is happening and how their characters are behaving and feeling.

The children use the medium of light to develop their individual interests and ways of learning.

On November 5th, a group of four year olds create a small world play scenario on a light box. This takes a considerable amount of time as the surface of the light box is covered, a bonfire is created and the characters are set out around the bonfire. The children have free access to:

- sheets and strips of coloured acetate
- ribbons
- different types of mesh.

The practitioner observes what the children are doing and listens carefully to how they make their decisions and what their thoughts are about the scene they are creating.

The scene created by the children expresses many of their thoughts, feelings and attitudes:

- the attendees at the bonfire are in family groups
- children are all accompanied by adults
- everyone is standing at a safe distance from the fire
- some children are standing with their backs to the fire.

The practitioner has the opportunity to interact with the children, exploring ideas around health and safety and relationships as well as finding out about their potential fears and concerns about the fire.

Other things to try

- **Encourage the children to express awe and wonder about the world, providing them with beautiful objects that prompt aesthetic awareness and an appreciation of things of beauty.**
- **Provide resources with a theme, such as sea creatures and blue and green materials, for the children to create new worlds.**
- **Use the light box to recreate stories which are familiar to the children.**
- **Enjoy playing with the light box and overhead projector yourself.**

> "Using the medium of light, children develop their individual identities and ways of learning"

> "Value children's ideas and ways of doing things"

Exploring shadows together

In the EYFS

The following statements are taken from the Practice Guidance for the EYFS, Personal Social and Emotional Development (Making Relationships).

- **Provide activities that involve turn-taking and sharing.**
- **Note the strategies that children use to join in play with individuals or groups of children.**
- ELG **Work as part of a group or class, taking turns and sharing fairly, understanding that there needs to be agreed values and codes of behaviour for groups of people, including adults and children, to work together harmoniously.**

Starting points

The three year olds in a nursery have many opportunities to explore and investigate using an overhead projector. They create interesting arrangements of resources and materials in their 'studio' space, where they can choose freely from the range of objects available to them. These include:

- transparent and translucent objects
- clear containers
- natural materials
- ribbons
- fine fabrics.

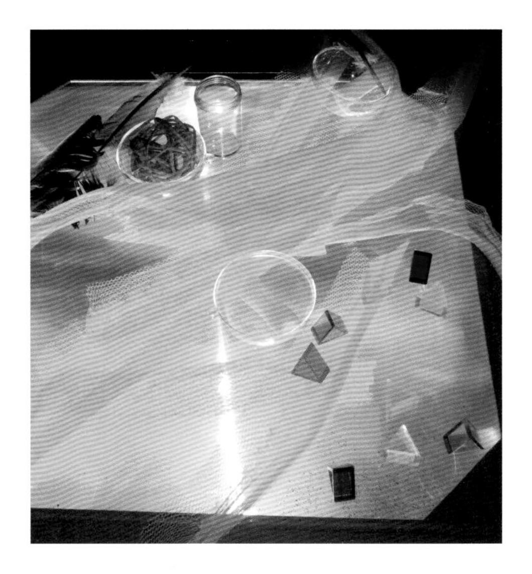

Learning and development

Three of the children create an arrangement on the overhead projector, taking turns and sharing the available resources. They move the resources around, explaining their personal preferences and accepting that their original choices might be changed.

They share the excitement of seeing the projected image of their arrangement on the wall . . .

. . . and rush over to explore it more closely.

While they are exploring the image they have created they realise that they are casting shadows of themselves on the wall.

They decide collectively to move the objects from the surface of the overhead projector so that they can explore their own images more easily.

As the three children play they are watched at a distance by a fourth child who can see that his shadow has joined the group!

He decides to approach the three protagonists to join in their play and the three happily involve the new member in their group, encouraging him to join in their collaboration.

Other things to try

- **Encourage pairs of children to play matching games using the overhead projector.**
- **Take photographs to show children how well they have explored together.**
- **Ask families to contribute special items to use with the overhead projector.**
- **Make opportunities for groups of children to explain what they have been doing, and what they have discovered, to a larger group or to other adults.**

Telling stories with shadows

In the EYFS

The following statements are taken from the Practice Guidance for the EYFS, Communication, Language and Literacy (Language for Communication).

- **Foster children's enjoyment of spoken and written language by providing interesting and stimulating play opportunities.**
- **(Children) consistently develop a simple story, explanation or line of questioning.**

Starting points

The three and four year olds in a rural children's centre are regularly visited by a visual artist, a musician and a story teller as part of a local arts project. The artists use light and shadow as a way of engaging the children's interest and bringing the stories to life.

The children benefit from having opportunities to retell stories using large scale shadow puppets and soft toys as story props. The artists bring with them a portable, folding screen and a powerful lamp to create the shadow theatre.

Learning and development

The artists help the practitioners to foster the children's spoken language by providing an interesting and stimulating environment which encourages the children to retell stories, build their vocabulary and think about how the characters feel.

The artists and the children build up a story of the farmer sowing his seed, the birds visiting the cornfield, the scarecrow scaring the birds and the corn growing successfully. The performance is enhanced by the musician and the children singing 'I'm a dingle dangle scarecrow'.

At the end of the performance with the shadow puppets, the children decide collectively that they want to change the ending of the story so that the characters can celebrate the harvest together. They include their soft toys which have become part of the performance.

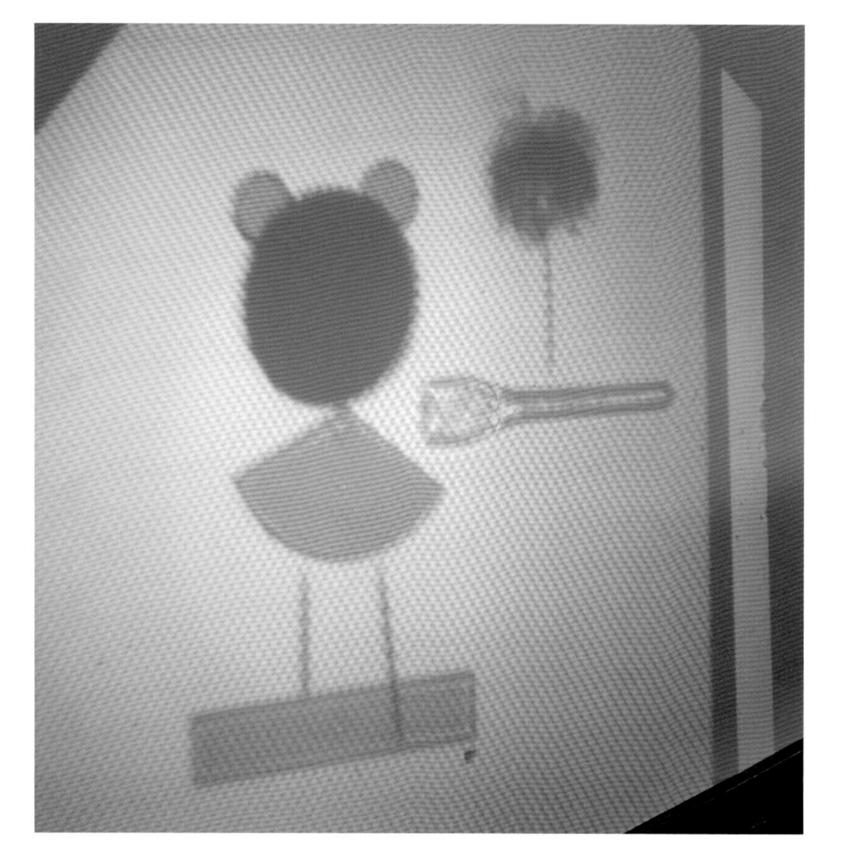

Two of the four year olds make connections in their learning by deciding to carry on the theme of storytelling using the overhead projector. They begin to design characters and a 'set' using open-ended materials including:

- plastic shapes
- glass beads and buttons
- an ice cream spatula
- a clematis seed head.

They cover the surface with a wire mesh tray to give a hazy effect.

The children then change their idea and rearrange the resources on the overhead projector to make a palm tree on an island in the bright sunshine.

Small acetate folders are added to make the sea.

Finally the children spend a long time carefully straightening two metal paper clips to add a seagull and a cloud to their picture. As they work, the children tell stories about the images which they change and project on to a wall.

Other things to try

- **Use a white sheet in place of a screen for successful shadow play - remember to use an area of your setting where the light can be excluded.**
- **Cut out characters from birthday cards to use on the overhead projector in story making.**
- **Natural materials such as skeleton leaves, seed heads, cones and pods create imaginative landscapes.**
- **Lacy fabrics create interesting backgrounds for storytelling.**

The wonder of light and colour

In the EYFS

The following statements are taken from the Practice Guidance for the EYFS, Communication, Language and Literacy (Language for Thinking).

- Create an environment which invites responses from babies and adults, for example, touching, smiling, smelling, feeling, listening, exploring describing and sharing.
- Be aware that young children's understanding is much greater than their abilities to express their thoughts and ideas.
- (Babies) are intrigued by novelty, and events and actions around them.

Starting points

Babies are intrigued by new experiences, events and actions around them. They show interest in their surroundings and respond to interesting stimuli.

In a day nursery, the Baby Room has a large light box placed on the floor so that the babies can explore unusual resources in a focussed environment which the light box creates.

The practitioners have created framed panels made of coloured acetate for the babies to investigate. A piece of wood has also been placed on the light box as a contrast to the coloured panels.

Learning and development

An eight month old baby notices the coloured panels on the light box and pulls himself up so that he can investigate them more fully. Holding on to the toy he has been playing with, he gazes intently at the nearest panel which has caught his interest.

Two of the older babies become interested in the panels on the light box and begin to explore the panels, picking them up and moving them around. The first baby's toy is abandoned as he concentrates on what is happening.

Non-verbal conversations are taking place between the three children which the practitioner observes and photographs. This will enable her to interpret, and give meaning to, the babies' investigations.

The youngest baby continues to watch intently as the others use their physical skills to reach and move the panels and to examine them more closely.

The babies are involved in active learning, remaining engaged and involved for a sustained period of time.

Other things to try

- **Provide the babies with coloured transparent objects such as picnic bowls, plates and beakers to explore using the light box.**
- ***Rainbow Sound Blocks* will combine exploration of light and sound using the light box.**
- **Sheets of coloured acetate and colour paddles can be used to introduce colour mixing.**
- **Try blowing bubbles so that they land on the surface of the light box – intriguing for babies to see.**

Exciting ways to sort and count

In the EYFS

The following statements are taken from the Practice Guidance for the EYFS, Problem Solving, Reasoning and Numeracy (Numbers as Labels and for Counting).

- **Provide collections of objects that can be sorted and matched in different ways.**
- **Provide collections of interesting things for children to sort, order, count and label in their play.**
- **(Children) Show curiosity about numbers by offering comments or asking questions.**
- **Use some number names and number language spontaneously.**

Starting points

Using open-ended resources for counting and sorting, rather than objects which are bought specifically for that purpose, encourages children to play with their mathematical ideas and think critically. It is important to remember that being creative involves all areas of learning with Problem Solving, Reasoning and Numeracy, providing many opportunities for creativity and critical thinking.

Children will be able to make connections in their learning if the environment encourages them to do so. In a nursery transparent, translucent and opaque materials are provided for the children to sort and count. They are stored in transparent pots made of thick glass and placed on a *Lazy Suzy* for easy access. A variety of containers are available for the children to use when sorting and counting.

Learning and development

A two year old is intrigued by the open ended resources which are available for her to explore on the light box. She spends half an hour absorbed in placing and moving the resources.

She carefully sorts the objects on the surface of the light box:

- linking people
- counters
- magnetic balls
- glitter buttons
- ice cream spatulas
- shiny pebbles.

The resources which she perceives as natural resources are carefully removed and placed on the corner of the light box.

The two year old carefully encloses the groups of resources which she has sorted by forming a ring using small acetate card covers. The practitioner is able to identify the child's learning style and records the learning process which she is going through.

There are clear links demonstrated between Problem Solving, Reasoning and Numeracy and Exploration and Investigation (Knowledge and Understanding of the World): *Encourage young children as they explore particular patterns of thought or movement, sometimes referred to as schemas.*

A three year old explores the resources on a light box, sorting and counting the objects in turn. She begins by removing the natural materials from her group of objects, and then starts to sort the remaining objects by colour. She becomes particularly involved in separating out the clear plastic shells, sorting them into matching pairs and developing her ability to count up to 8.

A four year old sorts the objects on the light box and counts the items in each group. This gives him the opportunity to count up to 10 objects.

By placing the triptych mirror on the light box, the practitioner has created the opportunity for the children to count numbers greater than 10, to estimate how many objects they can see and to check by counting them. They could also play number games with the light box and mirror.

A rich variety of interesting resources on a light box encourages children to investigate numbers, to count, to order, to explore position and solve problems in their mathematical play.

Other things to try

- Place specific groups of objects on the light box to encourage counting - three shells, three skeleton leaves, three buttons, three glass nuggets etc.
- Vary the objects you provide - 1 fir cone, 2 pebbles, 3 transparent counters, 4 colour paddles, 5 linking people etc.
- Cut out numbers from coloured acetate and encourage the children to select the correct one to go with their groups of objects.
- Encourage pairs of children to create number games and to set mathematical problems for each other to solve.

Make it big!

In the EYFS

The following statements are taken from the Practice Guidance for the EYFS, Problem Solving, Reasoning and Numeracy (Shape, Space and Measures).

- **Plan opportunities for children to describe and compare shapes, measures and distances.**
- **(Children) Show interest in shape by ...talking about shapes or arrangements.**
- ELG **Use language such as 'circle' or 'bigger' to describe the shape and size of solids and flat shapes. Use everyday words to describe position.**

Starting points

In a day nursery children are encouraged to explore shapes and space by using a wide range of transparent, translucent and opaque objects on light boxes and the overhead projector. They are given the time and space to play and explore.

The children in the nursery are active learners with a high degree of independence and control over their learning. They have regular access to a space with low lighting and an overhead projector which they use to project images on to a light-coloured blank wall.

Learning and development

The three year olds have become very interested in exploring simple shapes. The practitioner has provided a limited selection of transparent *attribute* shapes and some skeleton leaves to encourage the children to explore shapes, size and patterns.

Two of the children decide that they would like to recreate the large shapes they have projected on to the wall on a piece of paper and they set about deciding how to solve their problem of fixing the paper in the right position. They use language of shape and height as they work.

The position of their own shadows leads to further discussion and they decide to use the play opportunity for a different purpose. They secure the paper to the wall and talk about how they will solve their practical problem. They discuss their ideas, drawing on their previous experiences.

The children agree a route forward together and the exploration of shape, space and measures continues.

The activity interests the rest of the three year olds who listen while the original two protagonists explain their investigations.

The youngest girl stays with the activity and encourages two friends to explore with her. Although the boy is keen to be drawn around, the girls are by now much more interested in drawing round the shapes which have been projected onto the wall.

The whole experience has provided a context for learning which will be used by the children as they develop their mathematical understanding. When children are encouraged to play with resources and equipment they are much more likely to be able to use them in the future to solve problems.

Other things to try

- Use translucent shapes to encourage children to select specific named shapes and project their images.
- Use the linking people to develop positional language.
- Encourage the children to make large scale patterns with objects of different shapes. Use buttons, lids, conkers, leaves and laces.
- Play a game with the children - project the images of shapes, cover then partially with opaque material and gradually reveal the shapes until the children can name them.

The world in colour

In the EYFS

The following statements are taken from the Practice Guidance for the EYFS, Knowledge and Understanding of the World (Exploration and Investigation).

- **Encourage children to speculate on the reasons why things happen or how things work.**
- ELG **Investigate objects and materials by using all of their senses as appropriate.**
- ELG **Ask questions about why things happen and how things work.'**

Starting points

From the earliest age, babies are naturally curious, eager to find out about the world and how it works. They need opportunities to explore objects and materials to investigate and manipulate.

A light box and *infinity mirror boxes* are a part of the everyday environment in the Baby Room of a day nursery. The babies explore appropriate resources on the light box. In this instance the practitioner has provided large pieces of coloured acetate for the babies to explore on the light box.

Learning and development

An eight month old baby is playing near the light box when she discovers the pieces of coloured acetate left on the box by the practitioner to encourage exploration and investigation. The baby reaches for the blue acetate pieces and looks at them intently.

She then explores the acetate pieces by putting them in her mouth and licking the smooth surface.

The practitioner moves close to the babies as they continue to explore the acetate shapes and talks to them about what is happening. The practitioner slips some green acetate on to the light box to extend the investigation. One of the babies drops the blue acetate on top of green and is excited by the result.

She enjoys watching to see where the acetate pieces land when they are dropped and continues to maintain her interest in the investigation for a lengthy period of time.

The practitioner continues to show a genuine interest in the baby's discoveries, building the foundations for sustained shared thinking between the adult and the child.

A three year old working at the light box in a children's centre is equally intrigued by what happens when you look through transparent material - a blue plastic shape which he has been using on the light box.

He then decides to explore what happens when he looks at the different objects on the light box through a sheet of yellow acetate, achieving a very different visual effect.

The boy then places the acetate on the surface of the light box and lays out natural materials on the coloured surface which he has created.

He then begins to make connections with his earlier learning, looking through the blue plastic to change the colours he sees.

Both children working at the light box have been so engrossed in their own exploration and investigation that they have worked companionably alongside one another for ten minutes, each making their own discoveries and establishing connections in their learning.

Other things to try

- Provide colour paddles for the children to look through and experience the effects of mixing colours.
- Give babies transparent bottles containing feathers, buttons or beads to explore.
- Large pieces of jewellery made of transparent materials - bangles, necklaces and hair decorations - are suitable for older babies and toddlers.
- Use watertight containers, or bottles, with coloured liquids inside on the light box. Try including glitter or tiny stars in the liquids.

“Give children opportunities to speculate on the reasons why things happen and how things work”

“Create an environment where children can make their own discoveries and establish connections in their learning”

Natural light

In the EYFS

The following statements are taken from the Practice Guidance for the EYFS, Knowledge and Understanding of the World (Exploration and Investigation).

- Provide opportunities to observe things closely through a variety of means, including magnifiers and photographs.
- ELG Find out about, and identify, some features of living things, objects and events they observe.
- ELG Look closely at similarities, differences, patterns and change.

Starting points

Using a light box enables practitioners to present a range of interesting resources which inspire children's curiosity and encourage them to look closely, focus their attention and pay attention to detail.

Objects which are placed on a light box take on a different appearance and are drawn into sharp focus. By providing a wide variety of resources and materials like the ones below, practitioners can encourage children to ask questions and develop their own ideas and theories about the world around them.

- transparent, translucent and opaque objects
- natural materials
- decorative items made of transparent materials

"Using a light box helps children to focus on the patterns and shapes of objects"

"Encourage children to use the correct scientific words – transparent, translucent, opaque"

"Look at similarities and differences, patterns and change"

Learning and development

The practitioners in a nursery set out a selection of natural resources on their light box and invite the four year olds to explore them. They set out:

- seed pods
- leaves
- shells
- sponge
- lavender
- feathers
- pebbles.

The light box is set up in a darkened area of the setting so that the children can focus on the patterns and shapes of the objects on the light box

The children are interested to discover that they can see the details of some of the translucent resources clearly. The opaque objects also look different and attract careful attention. The practitioners encourage the children to use the correct scientific vocabulary – transparent, translucent and opaque.

Some of the children are particularly interested in the skeleton leaves and they work together to create a display of leaves on the light box. They talk about the similarities and differences between the leaves, the patterns of the veins and how the leaves have changed into skeletons.

Using a light box enables children to look very carefully at a collection of leaves collected in the local area. They can see clearly the shapes, structures and colours of the leaves. The children become absorbed in their exploration of the leaves, investigating the many different shades of green they can see.

The practitioners provide a range of tools for the children to use alongside the light box:

- colour paddles to use for colour comparison
- box magnifiers and hand lenses to look closely at detail
- wooden miragescopes to give a kaleidoscope view of the leaves.

Other things to try

- **Packets of pot pourri contain interesting items to be examined on the light box; they also add the dimension of 'small'.**
- **X-rays used on a light box are a fascinating way for children to learn about parts of the body.**
- **High quality plastic mini beasts make a good alternative to live creatures for children to observe closely.**
- **Provide examples of the same object made from a variety of materials for the children to observe differences. For example, pebbles made of stone, glass or wood will all look very different.**

> "Interesting resources inspire curiosity and encourage close observation"

Getting it right with light

In the EYFS

The following statements are taken from the Practice Guidance for the EYFS, Physical Development (The indoor environment).

- **Physical development helps children to develop a positive sense of well-being.**
- **Provide equipment and resources that are sufficient, challenging and interesting and that can be used in a variety of ways...**
- **Encourage children to use the vocabulary of... feeling, such as 'excited', 'scared' and 'happy'.**

Starting points

Even for the youngest babies the indoor environment should provide a safe, secure but challenging space.

In a day nursery the Baby Room contains mirrored equipment and a large light box for the babies to access independently. The windows allow the babies to experience the natural light which is essential to their well-being, but the blinds enable the practitioners to change the light levels so that those babies who choose to do so can explore the light box.

Learning and development

Practitioners in two children's centres make good use of equipment and resources to encourage children to develop their physical skills, to develop a sense of wellbeing and to experience risky freedom.

The two and three year olds create a beautiful hanging by using their fine motor skills to:

- pick up and arrange materials
- cut and stick
- paint and mark make
- tear and twist.

The practitioners hang the banner in an area with plenty of natural light and the children are encouraged to observe the effects of the light at different times of the day.

A cosy space under the stairs is a favourite place for the three year olds to rest and play quietly. A small light box provides an interesting medium for the children to explore everyday resources, either individually or in pairs.

A simple box made of translucent material and containing a battery operated light source encourages children to move with control and co-ordination as they grasp the box and move it around to create exciting light effects.

The practitioners in a children's centre encourage the children to mark make using fine pens and clear paper on the light box. They document the children's physical development skills through photographs. They record the way the children develop the vocabulary of feeling by noting the words they use as they develop their fine motor skills.

'I think I can do my name on my own.' 'I can do lightning on my own. It's been lightning this morning. Last time when I was a baby I was scared of it. I won't be now if I'm three.' 'My name is a short name.'

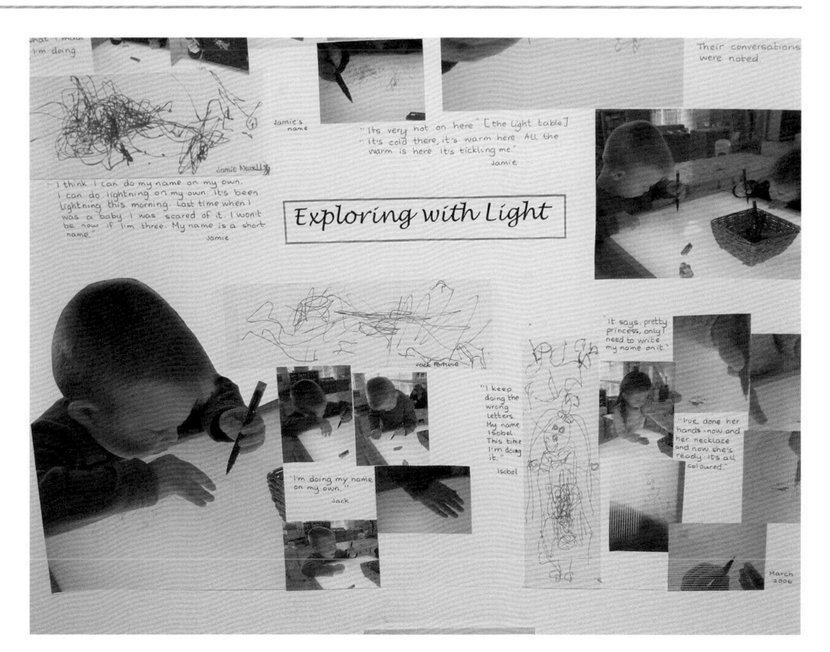

Other things to try

- **Make a space to explore light in your setting - try under a table, in a blanket tent or in a converted cupboard.**
- **Encourage the children to make large-scale movement using torches and fibre optic lights.**
- **Place a battery operated light in a shoe box, cover the shoe box with black paper (secure with a thick elastic band) and help the children to create light patterns and pictures by puncturing small holes in the paper.**

Shadows on the move

In the EYFS

The following statements are taken from the Practice Guidance for the EYFS, Physical Development (Movement and Space).

- Lead imaginative movement sessions based on children's current interests, such as space.
- Help children communicate through their bodies by encouraging expressive movement linked to their imaginative ideas.
- Demonstrate the control necessary to hold a shape or fixed position. Collaborate in devising and sharing tasks, including those which involve accepting rules.

Starting points

Children and adults alike are always fascinated by the effects of light and shadow.

Shadows are all around us but, as adults, we very rarely look at them.

Using the effects of light and shadow, both indoors and outside, encourages children to be aware of their own physical development, their own movements and use of space and their capacity to solve problems. Shadows are a wonderful resource for encouraging critical thinking and for making connections in learning.

The best thing about shadows is that they are a free resource!

Learning and development

On a sunny day the three year olds in a children's centre discover the joys of exploring their own shadows, making them move and stay still.

They spend time changing their positions and body shapes to see what happens.

They explore what happens to your shadow when you sit on it.

When the children have explored their shadows individually, the practitioners encourage them to make a group shadow, co-ordinating their movements and creating intended body movements.

The exploration of shadows is recorded in photographs and video to share with the children later in the day. There is no end product with an investigation of shadows!

When the three year olds in a nursery are painting with water out of doors, they take the creative step of deciding to paint their shadows. They soon realise what a challenge it is to paint something which is moving.

They decide it may be easier to paint each others' shadows.

The investigation lasts for over an hour as the children face the challenges raised by movement and evaporation!

Back indoors, the children transfer their knowledge and understanding of shadows by taking the puppets into the studio to use with the overhead projector.

The three year olds copy the movements and repeat the skills which they were practising out of doors, using their bodies to express themselves imaginatively.

Other things to try

- **Go on a shadow safari around your outdoor area to find different types of shadows.**
- **With the children, make shadow shapes with your hands - birds, insects, animals and monsters.**
- **Encourage the children to look at their shadows, indoors and out, as they jump, dance and balance.**
- **Encourage the children to see how big they can make their shadows.**

Mixing colours and light

In the EYFS

The following statements are taken from the Practice Guidance for the EYFS, Creative Development (Exploring Media and Materials).

- Provide a stimulating environment in which creativity, originality and expressiveness are valued.
- Ensure children feel secure enough to 'have a go', learn new things and be adventurous.
- Explore colour and begin to differentiate between colours.
- Differentiate marks and movements on paper.

Starting points

As children develop they explore a widening range of media and materials and begin to record their observations by mark making.

Three year olds in a nursery are accustomed to using light boxes and overhead projectors to explore and share their thoughts and ideas.

They have a wide range of materials and media available for them to choose from, including:

- transparent and translucent objects and containers
- opaque objects
- fabrics
- mark making equipment such as pencils, pens, pastels and chalks
- papers of different types, colours and sizes.

Learning and development

Two of the three year old girls in the nursery discover that they can explore individual colours using the overhead projector.

They choose a collection of red, yellow and orange objects and containers which they proceed to set out, and then pile up, on the surface of the overhead projector.

The girls talk about the colours they have chosen and the new colours they have made by 'mixing' materials.

They are very excited when they realise that they have changed the colour of the light in the room to a rosy glow.

The girls then begin to explore the effects of mixing other colours by overlapping translucent objects on the light box. They are delighted to show their explorations to the practitioners and the rest of the group of three year olds.

The children then capture their ideas about colour through representations, encouraging the other children to join them.

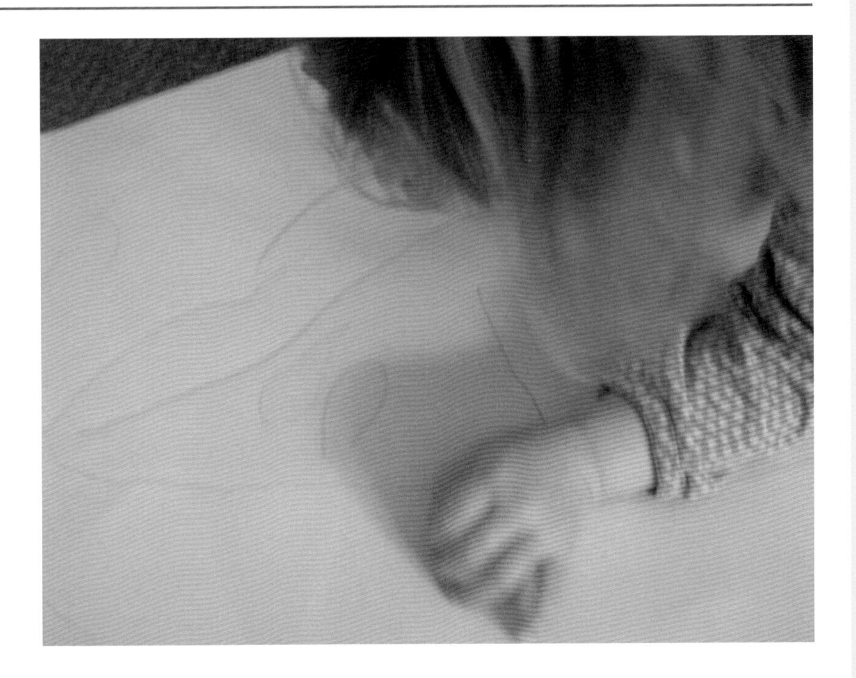

Other things to try

- Provide transparent and translucent resources in a single colour for investigation on the light box.
- Use sets of colour paddles in primary colours to encourage the children to mix colours on the light box.
- Give children a selection of clear reclaimed materials - small bottles, lids, containers, nozzles - to investigate on the light box.
- Provide younger children with sheets of coloured acetate to explore colour mixing.

A light gallery

In the EYFS

The following statements are taken from the Practice Guidance for the EYFS, Creative Development (Being Creative - Responding to Experiences, Expressing and Communicating Ideas).

- Encourage children to discuss and appreciate the beauty around them in nature and in the environment.
- Provide and organise resources and materials so that children can make their own choices in order to express their ideas.
- (Children) use language and other forms of communication to share the things they create, or to indicate personal satisfaction or frustration.
- (They) capture experiences and responses with music, dance, paint and other materials or words.

Starting points

Children respond in a variety of ways to what they see, hear, smell, touch and feel. As a result of different experiences, they express their ideas, thoughts and feelings in a variety of ways.

A day nursery has set aside space which is used by the children as a studio. There is plenty of natural light from the doors and windows which lead to the garden. The studio is equipped with a wide range of materials and tools to support Creative Development. These include:

- a range of papers
- a selection of paints and brushes
- water containers, including glass jars
- a light box
- an overhead projector.

Learning and development

The three year olds in the nursery have decided that they would like to create pictures of spring flowers. Three of the children, accompanied by the practitioners, visit the local shops and choose a bunch of spring flowers to paint. The flowers are put into a vase and set on a table in the studio.

The girls choose to use a palette of water colour paints and fine brushes to express their creativity.

Restful music is played in the studio as the girls work side by side, sharing the paints and taking turns as they clean their brushes in the jar of water.

The girls are encouraged by the practitioner to think about the beauty of the flowers and they respond with intense concentration as they portray the colours and shapes of the flowers.

The flower pictures are displayed on the light box, which the children feel makes them look even more beautiful.

The dirty water in the jar creates an effect of its own and the discussion moves on to what rain looks like.

The boys decide that they would like to paint the rain and they begin to use a new range of colours and shades and painting techniques.

The overhead projector attracts one of the boys and he stops to play with the translucent coloured resources en route to the drying area.

He then has a much better idea and decides to see what happens when his painting of rain is placed on the overhead projector. He uses his ICT skills to switch on the projector so that he can have a different perspective on his work.

When the rain paintings are placed on the light box they show clearly how the children have expressed their ideas about rain in a rich variety of ways.

Other things to try

- **Provide transparent paper or clear acetate for the children to draw on using felt tipped pens, resting on the light box. Do not have liquids on the light box.**
- **Drawings on acetate can be projected on to a wall forming large images which the children can interact with.**
- **The images can be projected on to large sheets of paper and traced over to produce large scale drawings and banners.**

USING MIRRORS

TEAM

Look at me!

In the EYFS

The following statements are taken from the Practice Guidance for the EYFS, Personal Social and Emotional Development (Dispositions and Attitudes).

- **Children must be provided with experiences and support that will help them to develop a positive sense of themselves and others. Providers must ensure support for children's emotional well being to help them to know themselves and what they can do.**
- **Place mirrors where babies can see their own reflections. Talk to them about what they see.**
- **(Babies) become aware of themselves as separate from others.**

Starting points

From a very early age, babies are fascinated by mirrors and by their own reflection. A child's first encounter with a mirror is an extraordinary event in their life. It is the first stage in being able to reflect upon themselves, both physically and metaphorically.

By placing a variety of mirrors – hand held, free standing or wall mounted – in an Early Years environment, practitioners can create many opportunities for children to develop a positive self-image.

Children will use mirrors as a starting point for active learning as they begin to make sense of the world around them and their place in it.

Learning and development

A mother and her five month old baby boy are taking part in a 'Stay and Play' session at a children's centre. A large *Mirror Exploratory* has been placed on the floor for the children and adults to explore. The baby shows an interest in the mirror and is supported by his mother as he begins tentatively to explore his mirrored surroundings. He is closely watched by a toddler, who peers over the top of the mirror.

The baby becomes more and more confident as he engages with his many reflections, gradually pulling away from his mother and supporting himself as he looks closely at himself in the mirror. He shows how deeply he is interested in his reflections by the length of time he concentrates on what he is seeing – over twenty minutes. He is beginning to see himself as an individual separate from others.

This experience provides the practitioners in the children's centre with a wonderful opportunity to talk to the baby's mother about his progress and development.

The babies in a day nursery benefit from having an environment that is rich in possibilities for exploration and investigation. A large *Triptych* mirror is often placed flat on the floor for the babies to crawl on to. They gaze intently at their reflections and begin playing with their fingers on the surface of the mirror.

When the baby is able to sit up unaided, the practitioner places a wicker ball within her reach. As she watches her own reflection and reaches for the ball to play with, the practitioner talks to her about what she can see:

'Look at you.'
'Who can you see?'
'What can you see?'

The older babies explore the mirror independently, spending time smiling, touching, gesturing and 'talking' to their reflections.

A mobile baby chooses to take with her a wicker ball from a treasure basket when she moves to explore the mirror. Although she is interested to see what the ball looks like on the surface of the mirror, she is more interested in exploring her own movements and expressions.

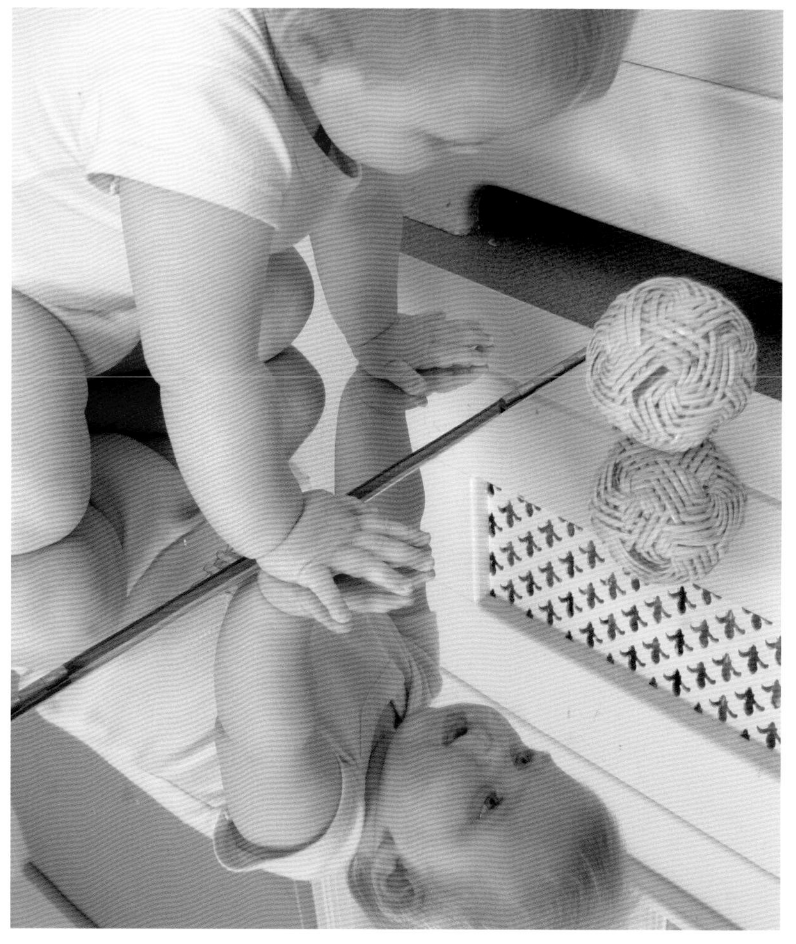

Large mirror cubes, *Mirror Exploratories* and hand-held mirrors continue to fascinate toddlers and older children as they develop their self-awareness and self-confidence.

In the toddler room of a day nursery, a two-year-old girl first places a toy inside a large mirror cube and then looks inside. She leans right inside and very quietly looks at herself and the toy from all angles.

She is very peaceful as she watches herself and is observed by two boys for some time.

When the little girl leaves to play elsewhere, one of the boys who was watching her decides to explore the mirror cube for himself.

He leans inside the cube and demonstrates a focused interest in what he is seeing. He spends almost ten minutes making eye contact with his reflection, looking intently at his own features and expressions. He is completely oblivious to the practitioner who is taking a photograph to share with him later in the day. This way, she can talk to him about what he was seeing and what he was feeling without interrupting his thoughtful self-reflection.

Other things to try

- **Place a mirror above, or to the side, of the nappy changing area in the baby room so that babies can see what is happening.**
- **Use message mirrors in your setting to help the children to develop their self-esteem. Use messages such as 'Today's special person is...'.**
- **Place mirrors in the quiet areas of your setting to create a reflective ambience.**
- **Wide mirrors, such as the *Mirror Exploratory*, allow adults to have conversations with children who find it difficult to make eye contact. As you talk to the child's reflection, you will be making eye contact with the child through the mirror.**

Creating a mirror den

In the EYFS

The following statements are taken from the Practice Guidance for the EYFS, Personal Social and Emotional Development (Making Relationships).

- **Create areas in which children can sit and chat with friends, such as a snug den.**
- **Provide time, space and materials for children to collaborate with one another in different ways, for example, building constructions.**
- **(Babies) seek to gain attention in a variety of ways, drawing others into social interaction.**

Starting points

Large *Kaleidoscope mirrors* serve many different purposes in an Early Years setting. They enable children of all ages to see themselves from different angles and to begin to know about themselves and the world around them.

A *Kaleidoscope mirror* is big enough to be shared by two or more children and can provide an interesting space for interaction with other children and adults.

In a day nursery, the *Kaleidoscope mirror* is used as a 'glass tent', beloved of the children in the infant toddler centres and pre-schools of Reggio Emilia in northern Italy. The babies and toddlers use the *Kaleidoscope mirror* to explore relationships and to create an area in which to be sociable.

Learning and development

The practitioners in the Baby Room of a day nursery have created a 'den' space using a *Kaleidoscope mirror* and some diaphanous fabric. A selection of interesting open-ended resources, including foam-backed, hand-held mirrors, have been placed inside for the children to explore.

The babies are encouraged to explore the space and the resources independently.

On other occasions, the light fabric is replaced by a blanket, which creates a darker space to explore. One of the babies tries to attract the attention of her reflection and shows 'herself' the block she is holding out. She is actively interacting with her reflection.

A second baby pulls the large soft caterpillar he is playing with into the *Kaleidoscope mirror* and shows the toy, its reflection. He shows his own reflection the rainbow sound blocks that he is shaking. He is developing relationships with himself and the world around him.

The sound blocks are picked up by a third baby later in the morning. He sits up straight, makes eye contact with the practitioner and offers up the sound block in an invitation to interact. The practitioner takes the opportunity to follow the baby's lead, sitting by the edge of the *Kaleidoscope mirror*, shaking a sound block and talking to the baby about what he is doing and what he can see.

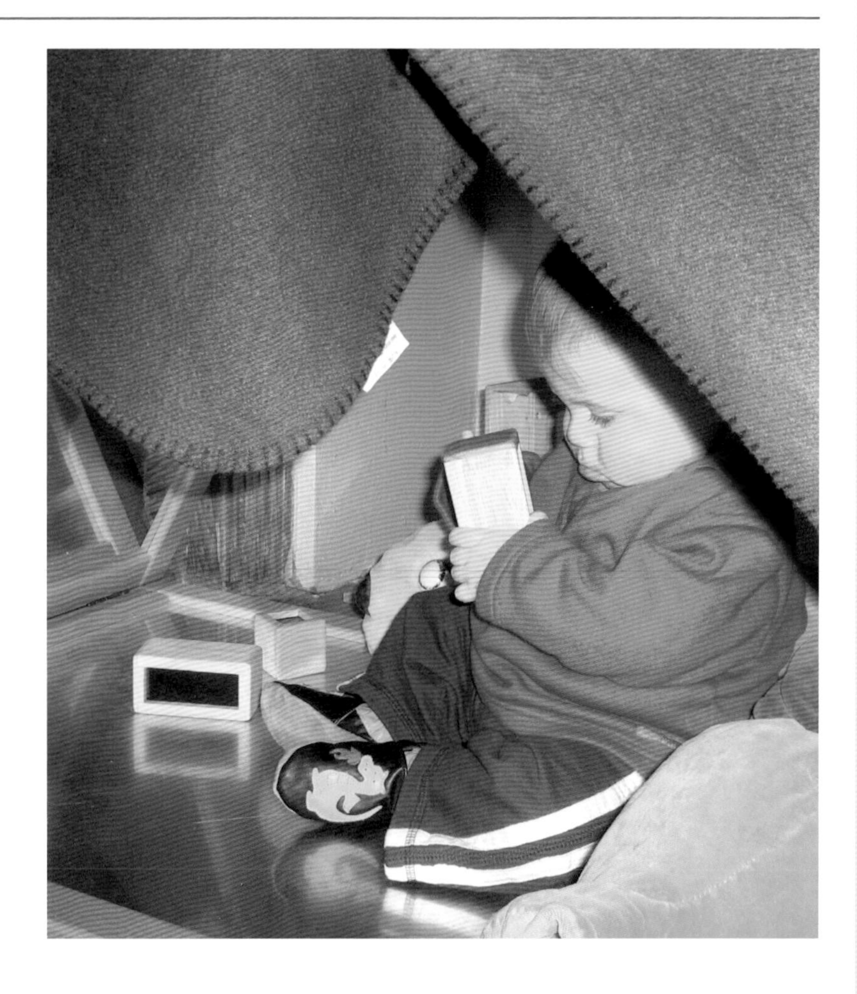

When the practitioner moves away, one of the mobile babies crawls into the *Kaleidoscope* den and begins to play companionably with her friend. The blanket is pulled further down over the mirror and the children's privacy is respected as they play together.

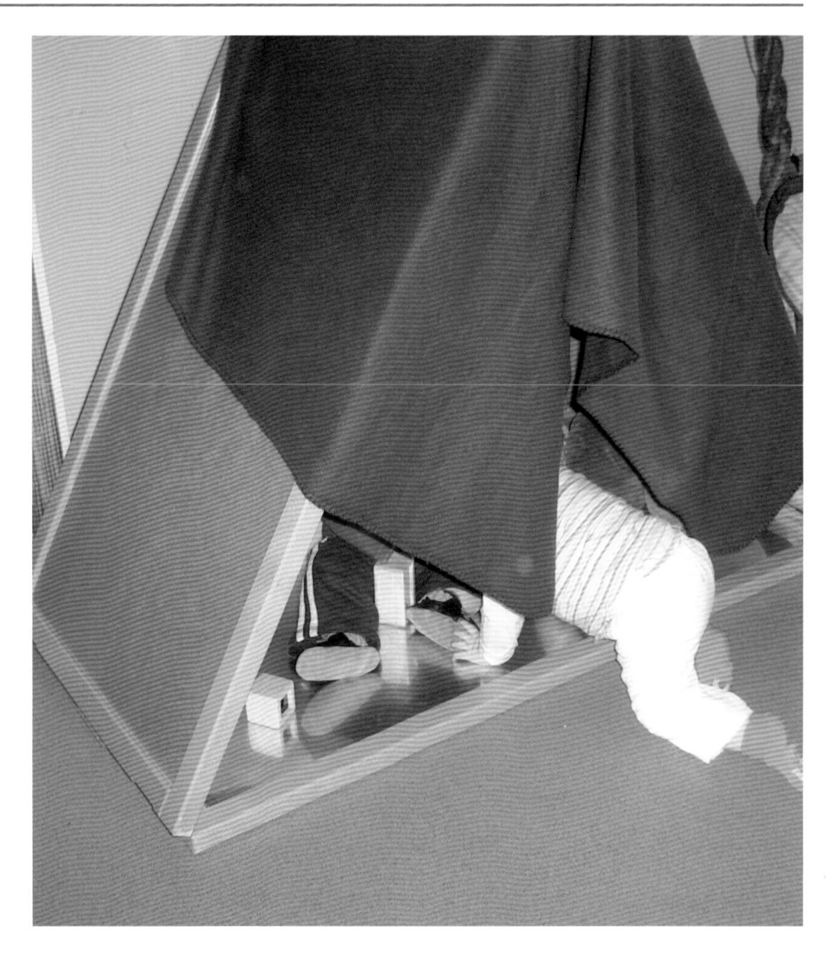

After a while, the babies try to attract the attention of the adults in the room by using their newly found physical skills and their voices.

When the toddlers see how the *Kaleidoscope mirror* is being used as a den in the baby room, they ask if they too can play with their friends in the new special place.

By changing the fabrics and resources available, the practitioners encourage the children to collaborate in using the *Kaleidoscope mirror* as a magical place for imaginative play with their friends.

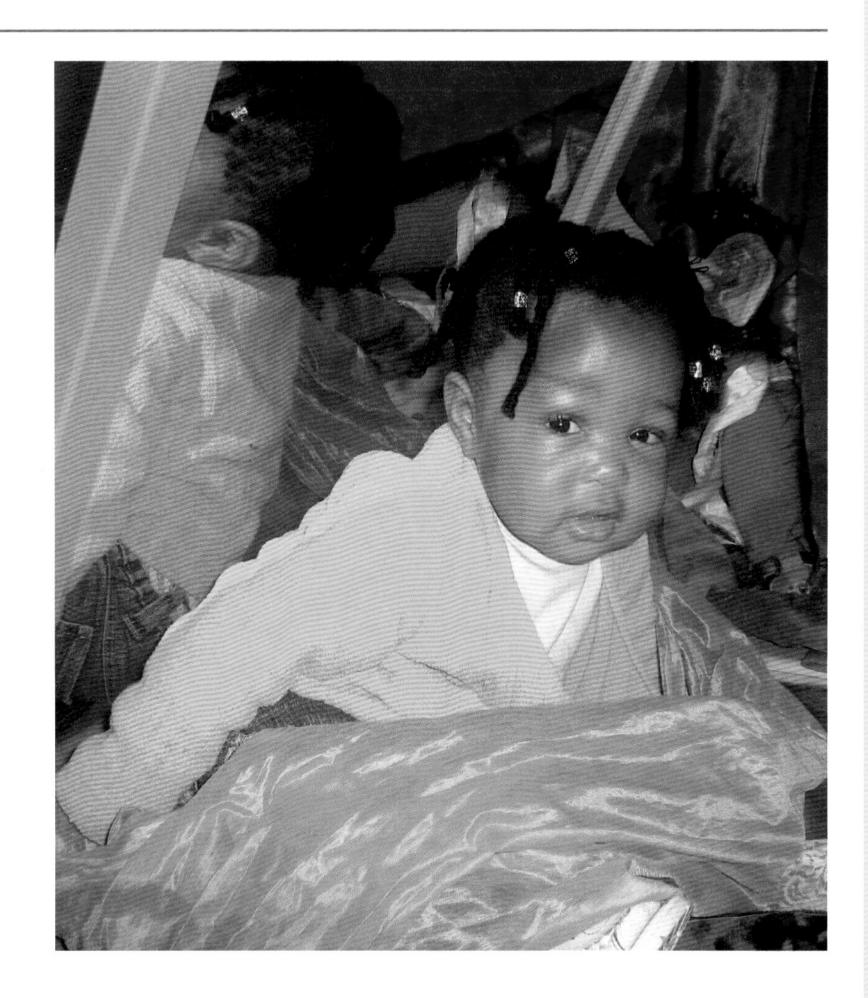

Other things to try

- **Sit inside the *Kaleidoscope mirror*, spending time talking to a child about who and what they see.**
- **If there are siblings in your setting, encourage them to sit inside the mirror together. Talk to them about similarities and differences. Try doing the same thing with children of similar ages or different ages.**
- **Use the *Kaleidoscope mirror* as a quiet zone as part of your behaviour-management strategy.**
- **Encourage parents to spend time inside the *Kaleidoscope mirror* with their children.**

“Mirrors help children to begin to know themselves and the world around them”

“A *Kaleidoscope mirror* creates a place where babies and toddlers can socialise and explore relationships”

Small world play stories

In the EYFS

The following statements are taken from the Practice Guidance for the EYFS, Communication, Language and Literacy (Language for Communication).

- As children develop speaking and listening skills they build the foundations for literacy, for making sense of visual and verbal signs and ultimately for reading and writing. Children need varied opportunities to interact with others and to use a wide variety of resources for expressing their understanding.
- (Children) use vocabulary focused on objects and people that are of particular importance to them.
- ELG Interact with others, negotiating plans and activities, and taking turns in conversation.
- Describe main story settings, events and principal characters.

Starting points

In a pre-school, the children have free access to a *Mirror Exploratory*, which they often use as an environment for small world play. The *Mirror Exploratory* enables small groups of children to play together, becoming absorbed in imaginative play for sustained periods of time.

The children regularly set up collaborative play situations, which result in group story making. As they interact with other members of the group, the individuals negotiate the story line, make plans about what they need to do and take turns in the conversation.

Learning and development

A group of boys decide that they want to create an imaginary hillside scene where they can create, and act out, a story using the toys that are currently popular in the pre-school.

They begin by selecting the materials they need to make the landscape for the story. They choose sheets of tissue paper and create hills, fields, paths and rivers with the green, brown and blue tissue.

The boys negotiate exactly where they will place the paper inside the mirrors to create the best effect.

The boys line up the toy dogs, bears and people on each side of the environment they have created in an adversarial setting. They tell the story of the two armies, giving reasons from the past for why they are enemies and what they are about to do. They develop a complex story line and introduce characters based on adventure stories they have heard or seen.

As the story unfolds, the boys realise that they have created a vast landscape by using the multiple reflections of the *Mirror Exploratory*. One of the boys remarks that the hills and rivers 'go on for hours!' The story becomes more complex as the characters travel across seas, deserts, mountains and forests in their quest for treasure. The boys are using their critical thinking skills to decide which way to move the characters

While the boys are engrossed in their story making, they are watched closely by a small group of girls who decide to build on the theme of story telling once the boys have moved on to act out their story elsewhere.

They carefully remove the characters and line them up on the outside of the *Mirror Exploratory* while they adjust the scenery to suit their purposes. As they do so, they talk to the characters and show them their reflections in the mirrors.

The girls create spaces that the families of characters can live in. They tell a story of families travelling from place to place, hiding and searching for somewhere they can be safe and secure.

Whilst the boys develop a story line of risk, challenge and adventure, the story making by the girls is built around family relationships and daily routines.

After a while, one of the girls takes the story making in a different direction when she says:

'Let's imagine it might be magic here and they have all disappeared. Abracadabra!'

She closes her eyes and uses gestures associated with magicians, covers the family up and makes them 'disappear'.

Her friend continues her part of the story line as they play together collaboratively.

Throughout the morning, the practitioner observes what the children are doing, notes down what they are saying and takes photographs. Later in the day, she encourages the children to revisit the stories they told and helps them to create a book telling the stories in photographs and their own words.

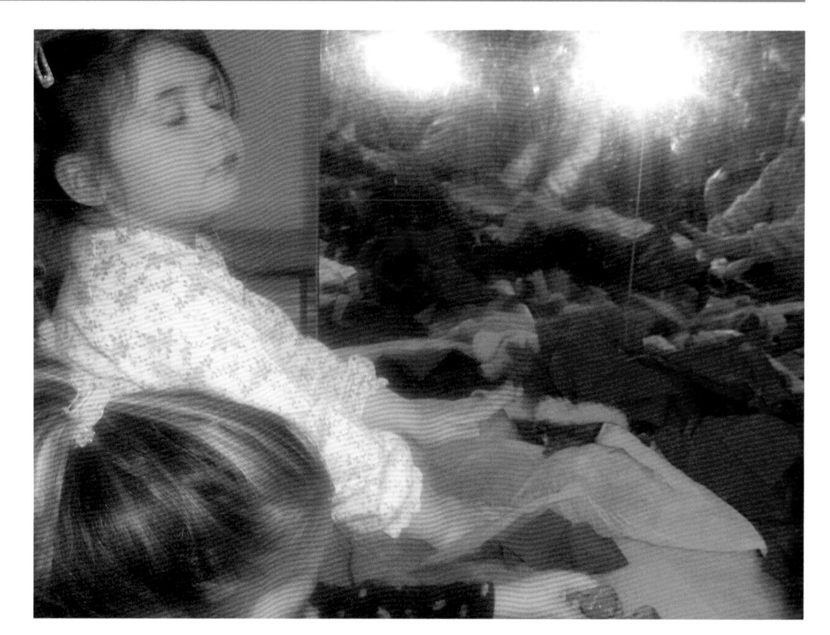

Other things to try

- **Provide a variety of papers, fabrics or bubble wrap to create imaginary scenes.**
- **Vary the characters available to the children – provide sea creatures, dinosaurs or farm animals.**
- **In conjunction with an acrylic mirrored environment, provide soft foam wood-effect blocks to create cities when reflected.**
- **Talk to the children about how they could use the mirrored environment to make a road layout with signs and symbols in the right places.**

First impressions

In the EYFS

The following statements are taken from the Practice Guidance for the EYFS, Communication, Language and Literacy (Language for Thinking).

- **Include things which excite young children's curiosity.**
- **Take an interest in what and how children think and not just what they know.**
- **(Children) use action, sometimes with limited talk that is largely concerned with the here and now.**
- **ELG (Children) use talk to organise, sequence and clarify thinking, ideas, feelings and events.**

Starting points

Young children develop their creativity and critical thinking skills through opportunities that help them to ask questions, pose and solve problems, and make connections within their learning.

In a children's centre, great use is made of mirrors that provide interesting and unusual views of the world, which arouse children's curiosity and challenge their thinking. A rich variety of mirrors that produce multiple images provide an opportunity for practitioners to discuss what is real and what is not real with the children.

Learning and development

A two year old girl looks inside a large mirror cube and explores the sides, touching the mirror and her reflection, and changing her facial expressions to see the resulting reflections. At this stage, she explores the mirrors without speaking, focusing on what she can see and the effects she can have on the reflections.

The two year old child rehearses an increasingly complex range of gestures and expressions, and observes herself from all angles, looking straight ahead and to the sides. She makes different sounds and looks closely at the shape of her mouth as she does so.

After a period of time, the girl notices her reflection at the base of the mirror cube. She decides to look more closely at the reflective image she can see. She puts her face close to the mirror and looks at the multiple reflections that appear to be below her.

The two year old is clearly using critical thinking skills, posing challenges for herself as she playfully explores the mirrors. She becomes fascinated by what is happening underneath the mirror and talks to the practitioner about what she can see and what might be hiding under the table.

The older toddlers in the children's centre are fascinated by fibre optic lights. They ask the practitioner to help them to position the lights in the mirror cube so that they can see the multiple reflections.

The three year olds build on their previous experiences as they play together with natural resources and small foam mirrors in a Mirror Exploratory. They talk to one another as they arrange the resources, connecting their ideas, explaining to each other what is happening and anticipating what might happen if they move the resources around.

The practitioner extends the children's thinking and language by talking to them about what they have been doing and helping them to reflect upon, and explain, events.

The older children in the centre's pre-school room are excited by the arrival of a full-length concave/convex mirror.

One of the girls looks at her reflection in both sides of the concave/convex mirror and talks to a watching practitioner about what she sees.

She uses language to describe shape, size, position and movement to express her ideas and theories. She talks about what she looks like and how she feels as she looks in the two different mirrors. The practitioner joins in the fun, sharing his own thoughts about how he feels when looking at the distorted images of himself. They play together, pulling faces and making different shapes with their bodies.

Other things to try

- **Make a mirror box with a shoe box lined with good quality acrylic mirrors and encourage the children to talk about what they are doing and what they observe as they place objects in their box.**
- **Display unusual items in small mirror cubes so that they can be seen from all sides - from above and below. Encourage the children to organise their thoughts and ideas as they explain what they are seeing.**
- **Use hand-held mirrors to see behind things and places. Ask the children to predict what they might see.**
- **Introduce the children to concave and convex mirrors from an early age. Talk to the children about the unusual images they see. Have fun!**

Solving problems with mirrors

In the EYFS

The following statements are taken from the Practice Guidance for the EYFS, Problem Solving, Reasoning and Numeracy (Calculating).

- Take note of the strategies children use to show they are working out if a group of objects is the same or different.
- Provide a wide range of number resources and encourage children to be creative in thinking up problems and solutions in all areas of learning.
- (Children) recognise groups with one, two or three objects.
- (They) use their own methods to work through a problem.

Starting points

In the pre-school of a day nursery, the children are encouraged to initiate their own learning experiences out of doors as well as indoors. They are accustomed to using open-ended, reclaimed resources in their play.

Mirrors and mirrored surfaces are fascinating to the children and they will spend a lot of time arranging and grouping resources such as buttons and glass nuggets in reclaimed trays and on mirror tiles, using their creative thinking skills as they sort, count and solve problems.

Learning and development

A pre-school practitioner is observing a group of three- and four-year-old boys as they investigate their chosen resources – reclaimed sectioned trays with shiny inserts, mirrored wooden blocks, glass nuggets and buttons – out of doors on a sunny day.

One of the boys is particularly interested in a reclaimed sectioned tray with removable mirrored pieces. He spends time moving the inserts from one place to another, counting the rows and columns that he is creating.

He then decides that he will place a glass nugget in each of the sections, demonstrating his understanding of one-to-one correspondence. He tells the practitioner that he needs the same number of nuggets as mirrored surfaces; he does not want to add nuggets to the holes in the tray. He poses problems to himself as he plays, talking out loud about whether he has enough, too few or too many glass nuggets to solve his problem.

A second boy who has been watching the problem raising and solving with interest moves closer and also uses a tray with mirrored sections to place and count the glass nuggets, following the lead of his friend.

One of the girls approaches and asks what is happening. The boys explain the strategies they are using to calculate the number of glass nuggets they need. She also learns from her peers and begins to place the nuggets on the mirrored inserts - one on each and then two on each mirror. Looking into the mirrors, she says:

'Look how many of me you can see.'

She then begins to count her reflections.

A younger boy has been playing alongside the others, placing the glass nuggets on the mirror tile. He is engrossed in watching himself as he plays. He begins to place the nuggets in the shape of his face, placing them directly onto his reflection.

He is closely watched by one of the older boys who asks him what he is doing. The three year old explains that he has been 'making his face' in the mirror.

The older child begins to create the image of himself he can see in the mirror tile using the glass nuggets. He says:

'Look, two for eyes,
one for a nose,
five for a smiley mouth.'

The observant practitioner has a photographic record of this playful mathematical encounter. She can use the evidence she has collected through the photographs and the words the children used to plan for the next stage in their learning.

Other things to try

- **Encourage very young children to look into a mirror and count with them - one nose, two eyes, two ears, ten fingers and so on.**
- **Use mirrors out of doors to look at what is above you and the children. Place the mirror on the ground and count the things you can see reflected.**
- **Label mirror tiles with numerals 1 to 5 or 1 to 10 and ask the children to place the appropriate number of things on each tile. Use cones, leaves, conkers or stones out of doors.**
- **Prop up a mirror next to where the children can arrange larger objects and talk to them about seeing double.**

"Children use creative thinking skills as they sort, count and solve problems"

"Use photographs and records of children's words to plan for the next stages in their learning"

Placing and arranging with a difference

In the EYFS

The following statements are taken from the Practice Guidance for the EYFS, Problem Solving, Reasoning and Numeracy (Shape, Space and Measures).

- Note how through talking about shapes and quantities, and developing appropriate vocabulary, children use their knowledge to develop ideas and to solve mathematical problems.
- Provide materials and resources for children to observe and describe patterns in the indoor and outdoor environment.
- (Children) show an interest in shape and space by playing with shapes or making arrangements with objects.'

Starting points

In a day nursery pre-school room, the children are familiar with selecting their own resources to explore the mathematical concepts of shape, space and measures in their play. The resources are stored in a clear cantilever artist's box which enourages the children to sort and classify the resources when they return them to where they are stored.

As part of the nursery's core experiences for children, a light table has been set up with a table-top triptych mirror placed upon it. This adds an extra dimension to the possibilities for creative and critical thinking.

The children are encouraged to pose and solve mathematical problems and are given time to become engrossed in their investigations without being interrupted.

Learning and development

A four year old girl uses the translucent resources stored next to the light box to explore different shapes, sizes and patterns. She concentrates for an extended period of time, glancing up to see the reflections of the patterns she creates as they are reflected in the triptych mirror.

Later in the morning, one of the boys selects a variety of round, square and rectangular shapes, and creates patterns and arrangements on the surface of the light box. He watches his reflection in the mirror intently and uses what he sees to develop his spatial awareness as he positions the translucent shapes. He uses the rectanguar shapes to create boundaries for the patterns he creates and uses the mirror to develop his early understanding of symmetry.

The practitoner builds on the interest the children have shown in placing and arranging resources using mirrors by setting out the light box resources and a tray of glass nuggets, pebbles and stones alongside a selection of mirrors. She waits to see how the children respond to the resources she has provided.

One of the boys is attracted to the mirrors and the resources. He randomly places some of the glass nuggets and pebbles on the round mirror and sits back to reflect on what he has done.

As he puzzles over the arrangement, he is involved in clarifying his thoughts and developing ideas of what he might do next. There is no pressure of time for him to complete the arrangement.

He begins to move the largest stones to the edge of the mirror, counting both the stones and their reflections with amusement. He also arranges the blue glass nuggets into a triangular shape which he is proud to name.

The boy's mathematical investigations have been observed by other members of the group who ask if they can join in. They continue to separate the large and small pebbles, to create a pattern around the boundary and this time, the blue nuggets are arranged in a square.

The children talk about the space remaining on the mirror's surface and discuss how it might be covered. They use the language of shape and space as they work together.

They take turns in placing the resources, watching and listening to each other's ideas, and commenting on their methods and choices.

Other things to try

- **Use plain table mats next to an upright mirror for the children to make arrangements and patterns that they can then see reflected in the mirror.**
- **Place a mirror next to a simple arrangement of shapes and talk to the children about symmetry.**
- **Introduce buttons, tangrams and mosaic pieces to be used with mirrors of different shapes and sizes.**
- **Place a large mirror on a construction platform so that the children can use building blocks to solve problems in three dimensions.**

Exploring shiny things

In the EYFS

The following statements are taken from the Practice Guidance for the EYFS, Knowledge and Understanding of the World (Exploration and Investigation (shiny things)).

- Note children's actions and talk in response to what they find and the questions they ask.
- Note the ways in which children find out about things in the environment, for example, by handling something and looking at it closely.
- (Children) explore, play and seek meaning in their experiences.
- (They) show an interest in why things happen.

Starting points

Young children show a great deal of interest in what appear to adults to be very ordinary things. By providing them with collections of objects with similar properties, their enquiries can be focused on particular features or processes.

By introducing a collection of shiny objects for the two and three year olds to explore, the practitioners in a day nursery begin to interest the children in reflections.

Learning and development

The practitioners display a selection of shiny objects in a large basket and wait to see what the children choose to explore and investigate.

One of the boys uses a shiny platter as a base and carefully positions a sieve and a spectacles case on the base. He looks across at the girls who have chosen a pewter tea set to examine.

He spends time looking at the reflections made by the sieve and the spectacles case, moving them around deliberately on the shiny tray to see what happens to the reflections.

The shiny spoons soon become a focus of interest for the group. One of the girls begins to move the spectacles case off the tray using a spoon. This turns into a collaborative effort as the boy helps by using his spoon.

While two of the girls are playing with the tea set and spoons, one of them makes an exciting discovery. She finds that her reflection is very different depending on which side of the spoon she looks into.

This piece of information is shared amongst the children in the group and they investigate the concave and convex sides of the spoons, talking to each other about what they can see.

The following day, the practitioner exchanges the objects that are in the basket for a different range of shiny objects, building on the children's previous interests.

One of the boys is particularly interested in the shiny CD. He turns it around, looking at his own reflections as he does so. He talks out loud about what he sees and what he is going to do next.

He decides to create a small, darkened place next to the basket using a sheet of acrylic mirror. He then looks carefully at his reflection in the mirrored top of a tin, watching himself as he changes expression.

The practitioner has placed sheets of paper within the children's reach, encouraging them to record what they can see in the shiny surfaces. Whilst some of the children prefer to continue to investigate the objects by handling them and looking at them closely, others choose to represent the reflections they see in drawings.

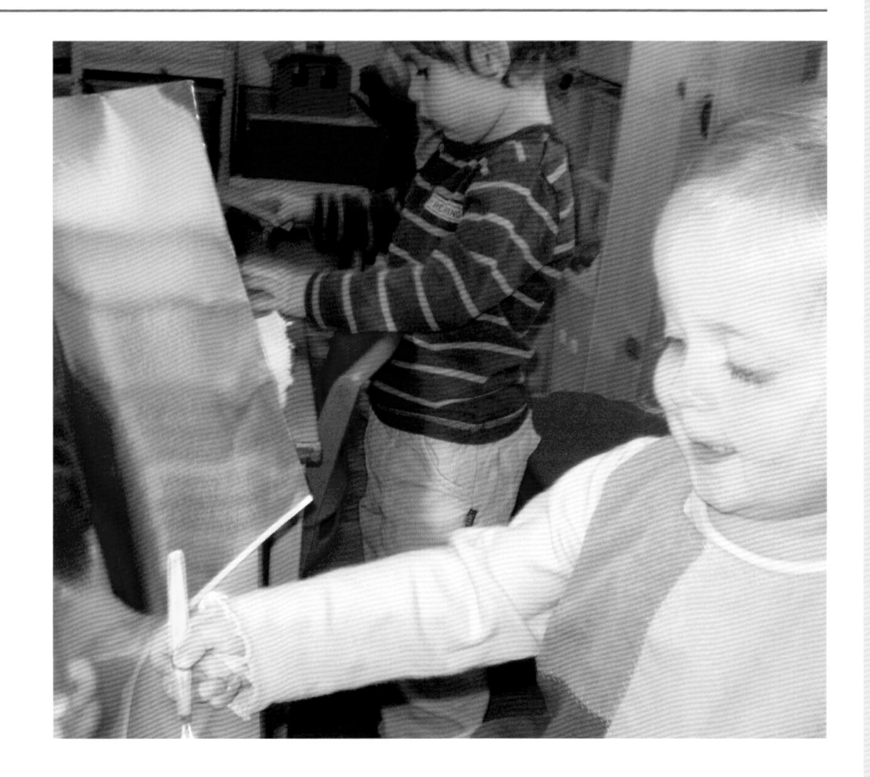

Other things to try

- Ask the children's parents to help you to build up a collection of shiny things for the children to explore. Encourage them to have a similar collection at home.
- When the children have investigated their reflections using shiny spoons, introduce small concave/convex mirrors to enable them to make connections in their learning.
- Encourage the children to look for reflective surfaces out of doors - in windows or puddles, for example.
- Talk to the children about the images they draw of themselves. This will help to build up an insight into how they perceive themselves.

“Young children show great interest in what appear to adults to be very ordinary things”

“Look for reflective surfaces out of doors – in windows and puddles”

Investigating light on the ceiling

In the EYFS

The following statements are taken from the Practice Guidance for the EYFS, Knowledge and Understanding of the World (Exploration and Investigation).

- **A rich and varied environment supports children's learning and development. It gives them the confidence to explore and learn in secure and safe, yet challenging, indoor and outdoor spaces.**
- **Encourage children to raise questions and suggest solutions and answers.**
- **Notice instances of children investigating every day events.**
- **(Children) explain their own knowledge and understanding, and ask appropriate questions of others.**

Starting points

Mirrors are a valuable addition to any early years learning environment. They provide intriguing viewpoints for children as they explore the world around them. A rich variety of safe acrylic mirrors are available that will enable practitioners to inspire young children's curiosity, creativity and critical thinking.

In a pre-school room, the practitioners make good use of mirrors to encourage the children to explore and investigate. A *Triptych mirror* is positioned on a light box so that the children can create patterns and pictures, and observe the multiple reflections. *Kaleidoscope viewers* are positioned to encourage the children to look closely at patterns, shapes, similarities and differences.

Translucent and opaque resources are often displayed alongside natural materials for the children to explore independently. Reclaimed resources are set out in a sectioned box and the practitioners have added two A4 mirrors to add interest to the way the resources are displayed. Hand-held plain mirrors and small concave/convex mirrors can be used to pose questions for the children to investigate further.

Learning and development

One of the four year olds has been playing with a hand-held mirror from the selection of mirrors available when he discovers that he can use the mirror to create an unusual effect on the ceiling by reflecting the sun.

His discovery begins to attract the interest of the other children in the room.

The boy explains to the others what he has done and they decide to use a range of mirrors to try the investigation for themselves. They watch each other carefully and talk about the different effects they are creating. They are very curious about what is happening and speculate about why the reflections are behaving as they are.

The children are excited by their investigative play and they begin to create reflections that interact with one another playfully.

They chase each other across the ceiling and the walls with their reflections.

The practitioner encourages the children to take photographs of the reflections they are creating to record their discoveries. She is then later able to talk to the children about what they have done and found out, encouraging them to ask questions about what was happening and to offer their own theories about what they have discovered.

Other things to try

- Place a variety of mirrors around your setting and talk to the children about what they can see at different times of the day.
- Use large combination concave/convex mirrors to see around corners, indoors and out of doors.
- Place mirrors on the floor or the ground outside and encourage the children to look at the reflections of trees, walls and the sky.
- When you are out for a walk with the children, take a selection of small acrylic mirrors, including plain and concave/convex, and encourage them to look at the reflections of things in the local environment.

"Mirrors inspire young children's curiosity, creativity and critical thinking"

"Encourage children to raise questions and suggest solutions"

Babies on the move

In the EYFS

The following statements are taken from the Practice Guidance for the EYFS, Physical Development (Movement and Space).

- **Support and encourage babies' drive to stand and walk.**
- **Offer low-level equipment so that babies can pull up to a standing position.**
- **Provide tunnels, slopes and low-level steps to stimulate and challenge toddlers.**
- **(Babies) make strong and purposeful movements, often moving from the position in which they are placed.**

Starting points

Practitioners can encourage babies to be involved in physical activity by using mirrors, which encourage them to make deliberate movements while sitting. Mirrors also foster babies' curiosity about themselves and the world, and prompt them to explore the space around them as they notice things about themselves and their environment.

Many baby rooms in day-care settings include a large triptych mirror that can be used either as a free-standing triptych mirror or laid flat on the floor. They have the advantage of being very stable and safe for babies to touch, press against and use to pull themselves up from the ground.

Learning and development

In the Baby Room of a day nursery, an eight-month-old baby sits in front of a large triptych mirror. She can see both herself and her key person reflected in the mirror. She is very excited by what she sees and spends time interacting with her reflection, pointing, waving, smiling and chuckling at the image of herself in the mirror.

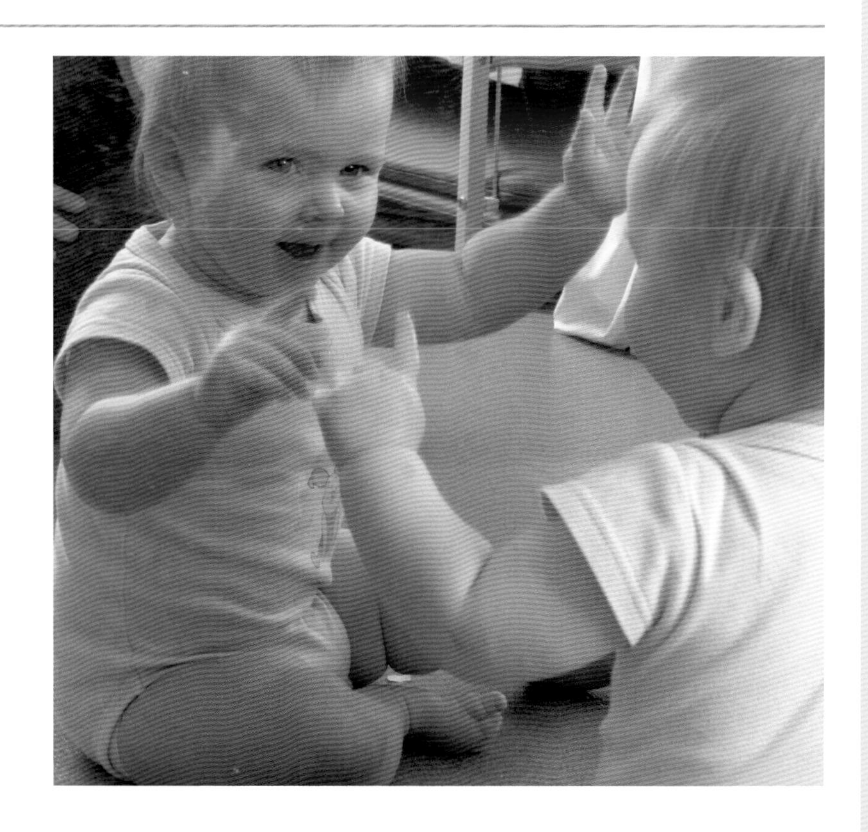

The baby is delighted by the effects her movements are having on her reflection. She changes sitting position and reaches up high with both arms stretching and waving above her head. She rocks gently from side to side as she waves.

The baby's actions become more and more animated and she makes contact with her reflection, co-ordinating the movement in both of her arms as she reaches out and up.

The practitioner watches intently but does not interfere in the baby's play as she connects with the image she sees in the mirror.

The baby becomes so excited by what she is seeing and doing that she sets herself the challenge of pulling herself up and standing in front of the mirror. She is delighted by her personal achievement and expresses her joy in the way she moves, smiles and laughs.

The baby's key person is clearly visible to the baby in the mirror. She leans forward and smiles, offering encouragement as the baby strives to stand up. By her gestures and facial expressions, she is showing the baby how delighted she is at what the baby has achieved.

On other occasions, the triptych mirror is set up in a corner of the baby room and a pop-up tunnel is placed so that it points towards the mirror.

The babies show great determination as they crawl and shuffle through the tunnel to reach the mirror. Because they are very familiar with mirror play, and enjoy interacting with their reflections, the babies rise to the challenge of finding different ways to travel through the tunnel. They watch themselves moving through the space, pausing occasionally to look up at their reflections.

The practitioners use the resources flexibly in the environment, moving them around to create new and exciting situations for the babies to experience, and building on the babies' interests and previous experience.

Other things to try

- **Sit in front of a mirror with a baby. Engage the baby in different physical experiences, such as rocking, bouncing, rolling and waving. Watch the baby's responses.**
- **Talk to the baby about the reflections you can see in the mirror. Encourage them to notice other babies and practitioners in the room.**
- **Introduce hand-held mirrors for the babies to grasp and handle.**
- **Place smaller acrylic mirrors in a line or circle on the floor. Observe what the babies do as they try to reach the mirrors.**

Reflecting on the outdoor environment

In the EYFS

The following statements are taken from the Practice Guidance for the EYFS, Physical Development (Movement and Space).

- **The physical development of babies and young children must be encouraged through the provision of opportunities for them to be active and interactive, and to improve their skills of co-ordination, control, manipulation and movement. They must be supported in using all of their senses to learn about the world around them and to make connections between new information and what they already know.**
- **Provide novelty in the environment that encourages babies to use all their senses and move indoors and outdoors.**
- **(Children) use movement and sensory exploration to link up with their immediate environment and show awareness of space, of themselves and of others.**
- **(They) respond to rhythm, music and story by means of gesture and movement.**

Starting points

Babies and toddlers are fascinated by mirrors when they encounter them. When mirrors are made available out of doors, a new dimension is added – the babies can see the wider world reflected in a large mirror.

A children's centre with limited space indoors for a sensory room created a 'sensory shed' in its garden. The smaller resources in the shed are changed on a monthly basis but the larger equipment, such as the triangular mirror, is used and stored in the shed on a permanent basis.

Learning and development

On a family fun day at the children's centre, the large triangular mirror is placed outside and the children and their family members are encouraged to play with it alongside other activities that are freely available. Having spent more than 15 minutes exploring the 'gloop' with his father, a nine month old baby boy is attracted to the triangular mirror.

His father holds him up against the mirror and he leans forward to look more closely at his reflection. At first he gestures to his reflection and then rests both hands on the mirror so that he can look directly at himself.

He moves backwards and forwards in a purposeful manner, watching the effects the movement has on his reflection. He begins to stand much more securely and confidently, and makes strong connections with his reflection in the mirror. He makes eye contact, leans forward on his arms and gently touches the mirror surface with his nose.

He shows great delight in his capability to stand almost unaided, leaning back confidently and supporting himself by pushing his hands against the mirror.

Throughout his time playing, the baby has continued to observe himself in the mirror, watching every move as it is reflected back to him. He has been supported, both physically and metaphorically, by an attentive, patient parent who is prepared to spend time encouraging his child to satisfy his curiosity about himself and his movements.

Inside the children's centre, two of the older toddlers are exploring reflections in a different way. They place a mirror on the floor in a spot where the sun is shining through the window.

Music is playing in the room and the toddlers use this as a starting point for their play. They select some cocktail stirrers and sit down to play together, negotiating who will sit where.

At first, they wave the cocktail stirrers over the mirror, taking turns and watching their reflections.

After playing together for several minutes, they hold a 'beater' in each hand and begin to respond to the rhythm of the music as it plays. They have become adept at negotiating the space and objects between them, and they play companionably together. As the music changes tempo, the toddlers make very intentional movements in line with the changing rhythm and beat. By using the mirror, they are very aware of their own movements and those of the other player.

By observing this child-initiated activity, the practitioner is able to give value to the growing awareness each of the children has of space and movement.

Other things to try

- **Position mirrors in the outdoor environment so that the children become more aware of the natural and built environments around them.**
- **Encourage the children to perform different actions in front of a mirror.**
- **Place mirrors where children can access musical instruments to encourage them to perform movements that are fast or slow, soft or strong.**
- **Encourage parents to use mirrors at home to help their children to develop a sense of space in relation to themselves and others.**

Catching shadows

In the EYFS

The following statements are taken from the Practice Guidance for the EYFS, Creative Development (Exploring Media and Materials).

- **Make time and space for children to express their curiosity and explore the environment using all of their senses.**
- **Note how children combine their creative skills and imagination to create something new.**
- **(Children) use ideas involving fitting, overlapping, in, out, enclosure, grids and sun-like shapes.**
- **(They) work creatively on a large or small scale.**

Starting points

Like artists, children choose to represent their ideas, thoughts, feelings and theories in many different ways. They will often choose unexpected ways to express themselves and practitioners should be prepared to build on the unexpected when it happens.

The pre-school children in a nursery are accustomed to developing their ideas, indoors and out of doors, in ways that interest them most. The practitioners give the children the time and opportunity to be creative with their ideas.

Learning and development

The four year olds in the nursery have been investigating their shadows out of doors on a sunny day. They have explored a variety of ways to 'capture' their shadows. Two of the children decide that they will chalk around the outlines of their shadows, talking about the difficulties they encounter as they work.

The rest of the group become interested in their idea and persuade the practitioner to lie down so that they can draw around her shadow. There is much discussion about the difficulties encountered in capturing shadows.

One of the boys is very sensitive to the health and safety aspects of facing the sun and he gently covers the practitioner's eyes with a piece of paper. One of the older boys brings along the sundial he has made 'just like Daddy's' and watches the shadows it makes.

The boy then returns to the pre-school room and returns a few minutes later with a hand-held mirror. He has clearly had a good idea that he wishes to pursue.

He begins to use the mirror to reflect the sun, forming spotlights on the ground. He then systematically moves around the outdoor area, creating spotlights to highlight each of the shadow drawings in turn.

He then develops his thoughts about using mirrors further by placing the mirror on the ground and looking to see what images he can create. A younger boy and a woodlouse are persuaded to look into the mirror and reflect upon what they can see!

The four year old asks the practitioner to take a photograph of him looking in the mirror so that he can have a record of his unusual viewpoint of the world.

This leads on to a week-long investigation of what the children see when they look up. The practitioner follows the children's ideas and provides the time, media and materials for them to represent their new views of the world. Eventually, a gallery display is created, which is shared with the children's families and visitors to the nursery.

Other things to try

- **Use stand mirrors to encourage the children to look closely at their faces. Encourage them to use fine pencils to draw individual features, paying close attention to the detail.**
- **Provide a range of resources for the children to arrange on mirrors. Suggest that they might like to represent the images they see using paints, pastels or crayons.**
- **Take the opportunity to develop the children's descriptive language as they work.**
- **Use a Mirror Exploratory to encourage the children to construct vertically and horizontally with blocks or small boxes.**

"Like artists, children choose to represent their ideas, thoughts, feelings and theories in many different ways"